Public Attitudes to Genetic Engineering:
Some European perspectives

EF/92/06/EN

European Foundation
for the Improvement of
Living and Working Conditions

Public Attitudes to Genetic Engineering:
Some European perspectives

by
Louis Lemkow

Universitat Autónoma de Barcelona
and
Instituto de Estudios Sociales Avanzados, Madrid

Loughlinstown House, Shankill, Co. Dublin, Ireland
Tel: +353 1 282 6888 Fax: +353 1 282 6456 Telex: 30726 EURF EI

Cataloguing data can be found at the end of this publication

Public Attitudes to Genetic Engineering:
Some European Perspectives

Luxembourg: Office of Official Publications of the European
            Communities

ISBN 92-826-4002-7

© THE EUROPEAN FOUNDATION FOR THE IMPROVEMENT OF
LIVING AND WORKING CONDITIONS, 1993. For rights of translation
or reproduction, applications should be made to the Director,
European Foundation for Improvement of Living and Working
Conditions, Loughlinstown, Shankill, County Dublin, Ireland.

Printed in Ireland

PREFACE

Over its last two four year work programmes, the Foundation has completed several projects which consider the implications of developments in biotechnology for living and working conditions. This work has brought together knowledge of the economic and social implications of the technology with the objective of supporting debate among the relevant actors about the application of biotechnology. It has become apparent that meaningful involvement requires a better understanding of needs and preferences for information of the different parties, particularly regarding genetic engineering as well as appropriate mechanisms for participation in the debate.

This report brings together results from two exercises which attempted to increase awareness of the attitudes and preferences of key parties included in discussions about biotechnology. The first project was based upon small focus groups with members of the 'informed public' to discuss their attitudes to various applications of biotechnology. The second project used a workshop format to establish priorities and concerns among specific interest groups, including trades unions, businesses, government, public interest, particularly environmental, groups and churches.

Both studies were done in the United Kingdom, France, Germany, and Spain. Funding for the work came from both the Foundation and Directorate - General XI (Environment) of the Commission of the European Communities. The Spanish study was supported by the Science Research Council, who also sponsored a meeting in Madrid to discuss preliminary results from the studies. Particular thanks are due to Professor Emilio Muniz, President of the Science Research Council and to Professor Salvador Giner, Director of the Institute of Advanced Social Studies, in Madrid. The author of this report that brings together results from the different studies is Louis Lemkow, a social scientist, who led the researchers in the Spanish study.

A review of existing studies of attitudes to biotechnology points to difficulties in their interpretation, and to the limitations of surveys in describing the key issues for the public and for interest groups. The relatively small qualitative studies presented in this volume complement results from other research and illuminate aspects of the concerns among different groups; they also identify strategies for addressing problems in information, participation and regulation. However, these studies also reflect the conceptual and methodological difficulties of this and other research; although they discuss specific developments in genetic engineering and offer a framework for organising debate about their implications. The report underlines the diversity of responses to genetic engineering among different population groups and between countries. It emphasises the importance of developing correspondingly diverse approaches to providing trusted information, and appropriate fora for public debate.

CONTENTS

INTRODUCTION	1
BACKGROUND	1
Medicine and Pharmaceuticals	3
Food Processing	4
Agriculture	5
Development and Dilemmas	5
PUBLIC ATTITUDES TO NEW BIOTECHNOLOGIES	8
PUBLIC OPINION SURVEYS	10
QUALITATIVE APPROACHES	17
Focus/Discussion Groups	18
The Sociopolitical Context	18
Knowledge of Biotechnology	19
Assessment of Benefits	20
Assessment of Risks and Safety	21
Access to Information	22
Regulation and Control	23
Attitudes and Diversity: Gender and Country Differences	24
The Workshops	25
Organisational and Technical Problems	25
Context	26
Issues and the Debate	27
Information	28
Decision-Making, Regulation and Control	29
Public Participation	29
Ethical Issues	30
Environmental Considerations	31
Socioeconomic Impacts	31
Health and Safety in the Workplace	32
Comparing and Contrasting the Focus Groups and the Workplace	32
THE CONTEXT AND IMPLICATIONS OF CONCERN	33
Agenda for Discussion	36
Future Research Needs	36
Information and Strategies for Participation	37
Regulation and Public Policy	38
BIBLIOGRAPHY	41

Introduction

The dramatic progress made in biotechnology in general and genetic engineering in particular over the past two decades is, in part, responsible for the new public debate on ethical, environmental, social and economic implications of man-made modifications to genetic material in micro organisms, plants and animals (including humankind). Attempts have been made by social scientists during this period to gauge public opinion in this polemical area of scientific research and development and today there is a substantial body of data on public attitudes to genetic engineering. Much of this research has been in the form of public opinion surveys. While a critical appraisal of this methodology will be presented here, the main purpose of this report is to outline the qualitative approaches used in the European Foundation's comparative study (Britain, France, Germany and Spain) on public attitudes to genetic engineering. The qualitative research options used in the Foundation's study are themselves not new: these were focus/discussion groups with what is termed the "informed public"; and workshops bringing together the different interested parties in genetic engineering. However, applying these methods to investigate attitudes to the new biotechnologies has complemented the existing survey research and has offered new insights into public concerns about current developments in genetic engineering.

Before embarking on an account of research into public attitudes to genetic engineering, the report considers some aspects of the development of biotechnology, especially in what is sometimes termed the "new" biotechnology. It aims to pinpoint areas of potential and real public concern in this rapidly expanding field.

Background

There are already exhaustive histories of biotechnology and its technical growth. This section does not summarise the major events that have taken place in this vast field but looks at how perceptions have shifted over time in relation to the development of specific applications. Biotechnology is frequently associated by the general public with the notion of new technology: it is seen as something pertaining to the scientific and technological advances made since the Second World War and largely concerned with genetic engineering and reproductive technologies. However biotechnologies are old indeed and go way back in history (and even prehistory). The brewing of beer, yogurt fermentation and wine making, all represent forms of biotechnology if we take a definition of these activities as "the application of scientific and engineering principles to the processing of materials by biological agents to produce goods and services" (Roberts, 1990).

The earliest biotechnologies were largely concerned with aspects of food processing and the transformation of certain raw foods, by biological agents such as microorganisms, into another edible form. Most of the public take these processes for granted considering them to be entirely natural. However, these processes involve human manipulation of a material through the controlled use of biological agents to create something different. This intervention may be considered as a form of technology or engineering, and may be labelled as biotechnology because the motor of such transformation is biological. In the past (as well as in the present) biotechnology played an important part in diversifying and enriching human nutrition. While these forms of food processing are known to everyone, they are not generally perceived as a threat or even as forming part of biotechnology. It is only the expert who refers to them as biotechnology. In recent times biotechnology has come to play an increasingly important economic role as an ever larger number of such products are developed and sold.

The nineteenth century saw dramatic advances in biotechnology, especially in the medical field. Vaccines against a wide variety of diseases began to be developed from viruses grown in the laboratory. At the same time biotechnology was used to treat and break down organic/domestic waste and effluent, playing a major role in achieving greatly improved levels of public health in the new industrial cities. The economic impact of these advances was considerable and resulted in the rapid development, marketing and sale of new and powerful products based on biotechnology.

The pace of biotechnology innovation and production increased rapidly after the Second World War especially in health care applications. This included the development of genetically engineered human hormones (insulin and human growth hormone), monoclonal antibody technology as the basis of new diagnostic products (markers for cancers and certain infectious diseases, pregnancy testing), and the polemical gene therapy.

In agriculture, developments have also been rapid - recombinant DNA has been used in the creation of genetically altered organisms to create new and more productive variants. Such innovations in agriculture, as with the case of some of the medical applications, have raised significant ethical, social and environmental questions. These have been reflected in increasing and intensive, though as yet not very extensive, public concern about developments in biotechnology. Public access to information, about both basic research as well as applications in the field, is one of the most clearly articulated demands emerging mainly from consumer and environmental groups but also from sections of the scientific community.

Medicine and Pharmaceuticals

The highly competitive and , in general, profitable pharmaceutical industry spends a very substantial percentage of its sales in research and development. This is especially the case with biotechnology applications, although it should be pointed out that many of the most technically advanced and innovative biotechnology firms are not responsible for the marketing and sale of their products.

One of the clearest examples of the potential impact of biotechnology is in the area of genetically engineered human hormones - particularly insulin in the treatment of diabetes. Until human insulin was first marketed in 1982, insulin was obtained from the pancreas of pigs and cows. Given that it was not identical to human insulin it sometimes resulted in undesirable side effects. There are other recombinant proteins used for medical purposes although as commercial ventures some of these have not proved successful. It is often the case in this type of biotechnology that the company undertaking the original R & D do not have the ability to be involved in the production and commercialisation of their innovation.

Home pregnancy test kits have become greatly simplified as a result of genetic engineering. Indeed a whole new generation of diagnostics has been produced - tests have been developed to identify and monitor certain types of tumours.

An increasingly important field of biotechnology research is concerned with the detection of genetic diseases. These DNA probes can be used to detect genetic diseases before birth; genetic information about the foetus can be used when deciding to continue or to terminate a pregnancy. Biotechnology has played an important role in advances made in a closely related field: that of "reproductive technology" (embryo freezing, **in vitro** fertilization, ovum implantation,) - an area popularly thought to be an intrinsic part of biotechnology. Developments in scientific research related to reproductive human biology have over a long period provoked considerable debate. Similarly, "gene therapy" involving the reprogramming of human cells through the manipulation of DNA has raised many ethical issues for scientists working in the field as well for the public in general. According to some critics such therapy is only a step away from the kind of genetic engineering envisaged in Huxley's "Brave New World" (Huxley, 1932) where human genetic material is used to create human beings with specific genetic traits.

Food Processing

An important volume of research and development in the food processing industry is linked to speeding up some processes such as fermentation. This is where new biotechnologies are coming to play an important role; for example, genetically engineered bacteria are used to quicken the maturation of cheese. Similar processes are involved in brewing and yogurt fermentation. An important consumer dimension is raised here through latent suspicion of genetically engineered food products. Resistance may be enhanced by apparently slight, yet for the consumer, significant changes in the taste or texture of the end product.

New foods can be developed by biotechnology. Certain soil fungi that can be grown on nutrients have been processed to create myco-proteins some of which, it is claimed, have the qualities of meat (though with less fat and more fibre and protein). Acceptability of new products is linked to information policy . There is, for instance, an ongoing and heated debate about the way in which biotechnology products should be labelled and evaluated (Jenkins et al., 1991).

Agriculture

In the agricultural sphere, biotechnology has been largely concerned with applications which will improve animal and plant growth and production. In relation to plant production (both for food and non-food use), biotechnology has been particularly concerned with the improvement of crop germplasm through the addition of new genes. Farmers and plant breeders have traditionally sought the "improvement" of plants for better performance in the field. There is a wide variety of conventional techniques for achieving specific improvements, the simplest being the crossing of plants (with certain desired characteristics) of the same species creating new combinations of the preexisting genes. Today, with the use of biotechnology, new crop species can be created by the insertion of a single foreign gene. This type of genetic modification was first achieved in 1982 and since that date progress has been very fast. There is now a lengthy list of transformed crops including wheat, rice, maize, potato, tomato, grapes and oranges. These are genetically engineered or modified organisms.

Developments and Dilemmas

The techniques are now available for transforming all kinds of organisms - from bacteria to animals. The examples outlined below have been chosen not only because they have received much attention by the media, but also because they were among the products presented in the focus/study groups of the research project (along with the already mentioned DNA probes).

BT - biopesticide: This is a genetically modified bacterium containing a gene from another one (**Bacillus thuringensis**) which enables it to kill insects, and can therefore be used as a pesticide. The modified living organism is mixed with water, and is spread on the plants to be protected (usually vegetables). When the insect eats the plant, it eats the biopesticide. The BT toxin attacks its digestive system; it stops eating and dies within a few days.

The BT bacterium, which has the toxin producing gene, occurs in nature and has already been used as a biopesticide. However, the BT bacterium does not survive well under agricultural conditions, is not very potent and only attacks a narrow range of insects. Genetic engineering removes the gene responsible for producing the toxin and puts it into another bacterium which survives better under agricultural conditions. This technology can also make different types of biopesticide attacking a wider range of insects. Many companies are currently carrying out research on BT-biopesticides, and field tests of this genetically modified organism are taking place.

BT-biopesticides raise a number of important issues about their impacts on environment and human health. It has been argued that the wider use of biopesticides can replace, in part, the use of chemical pesticides which pollute soil and water. At the same time it may be that these products are more specific and/or more effective than chemicals.

- The main environmental concern is that there is not enough knowledge about possible repercussions in nature and specific uses could upset the ecological balance in the following ways:
    * The bacteria may survive better than natural ones and out-compete them.
    * Beneficial insects may be more susceptible to the new more potent biopesticides, and a wider range of species may be threatened.
    * The toxin-producing gene may "jump" from its designed host to other bacteria for which it was not intended.
    * The removal of an insect (accidentally or intentionally) may create space in the system for another pest or may upset the food chain of other animals (the same argument goes for the use of all types of pesticides).
    * There may be longer term evolutionary effects.

As far as health is concerned it may be beneficial in that the introduction of biopesticides might mean less residue of chemical pesticides known to be harmful to human health. The principal concern is that the safety of the residue of the new bio-pesticides on food have not been fully evaluated (if indeed this is really possible). The BT gene does not so far appear to have any harmful effect on humans, but there is a wider question about the safety testing needed for such new products.

Herbicide-resistant plants (also known as "herbicide tolerant" plants). These are plants (arable crops, vegetables, flowers, trees) which have been genetically modified to be resistant to, or tolerate, chemical herbicides which would normally kill them. An example is a tomato plant, into which has been inserted a herbicide-resistant gene from a bacterium. This enables the plant to break down specific herbicides against which it has been made resistant, and therefore to be unaffected by this chemical while other plants (weeds) around it are killed. It will also be possible to plant herbicide-resistant crops in fields which have previously been sprayed with the herbicide and are contaminated.

At present, research is being carried out on something like 30 crop and forest tree species, modified to withstand otherwise lethal or damaging doses of herbicide. Seeds for herbicide-resistant plants are among the first new biotechnology products to be used in agriculture.

Environmental concerns include the following:

- Genes conferring herbicide resistance may be transferred, by pollination, to other plants which may make spread of the accidentally altered plants difficult to control.

- The introduction of the new herbicide-resistant plants may result in greater use of herbicides damaging to the environment (poisoning of wildlife and destruction of habitats for certain organisms).
- Health concerns are similar to those for BT-biopesticides.

There are other concerns about the ethics of patenting life forms which have been modified from wild species - species which do not have owners as such.

Genetically modified fish: Research is currently being carried out on how fish (e.g. salmon) bred in fish farms can be modified to grow faster, by the insertion of a gene producing growth hormone. There can also be alterations which affect toleration to stress and crowded space and resistance to the many diseases and parasites which flourish in fish farms. Delaying sexual maturity is another line of research which would result in greater size. Some of these traits were also being introduced by traditional breeding techniques, but genetic modification is much quicker and can introduce completely new traits.

The benefits for the fish farmer are obvious; increased profitability due to the increased size of the fish, less disease and better survival rates. Environmental concerns are as follows:

- The modified salmon can easily (inevitably) escape from the fish farm and breed with wild salmon. This may result in the disappearance of wild species and the loss of genetic diversity because:
  * If enough interbreeding takes place, one uniform salmon species may evolve some of the characteristics of the modified fish.
  * Through interbreeding, the wild salmon will acquire new characteristics and may lose their complex adaptability system that allows them to move out to sea and return to their spawning grounds. This may affect their survival rate and reproduction. The modified fish have higher feeding requirements and will compete for the available food in the environment. In the context of food scarcity, other wild species will diminish (this is the case for most farmed fish whether or not they are modified genetically).

BST - Bovine Somatotropin: The above examples have been chosen because they highlight many of the key public concerns about biotechnology; namely environmental, public health and economic issues. Ethical questions have also been raised about the perhaps best known and polemical case of the use of biotechnology in agriculture: BST. Bovine Somatotropin, which is manufactured with the use of genetically altered bacteria is injected into cows making them produce more milk for the same amount of food intake. Here the debate has been centred not only on public health aspects but also on the question of animal welfare (possible discomfort resulting from greater production) as well as the fact that there is already a surplus of milk in Europe.

Public Attitudes to New Biotechnologies

Over the past decade there has been research on the public's attitudes to biotechnology in several industrialised societies. Most of the studies undertaken to date have been public opinion surveys which reveal a rather complex and ambiguous set of attitudes towards biotechnology. In part this reflects the fact that the term "biotechnology" itself is problematic.

Biotechnology is used as an umbrella concept under which techniques and applications covering many fields are found. Even today, many scientific meetings held on biotechnology are prefaced by lengthy discussions on the provision of the most adequate definition of the subject. This state of affairs is exemplified by the coexistence of 41 different definitions of biotechnology in European Community documents. Indeed, as one historian of biotechnology has put it: "The number of definitions of the word biotechnology has become a matter of embarrassment" (Budd, 1989). The term biotechnology first came into use in 1919 when the Hungarian agricultural engineer Karl Ereki defined it as "all lines of work by which products are produced from raw materials with the aid of living organisms" (ibid.). Since that time a range of definitions has been developed, discussed, revised and argued about. It should come as no surprise to discover that the public in general is rather confused and unclear, if it has any views, as to what actually constitutes biotechnology.

Another problem when trying to evaluate the results of research on public attitudes to biotechnology is that many of the studies have asked rather broad questions about biotechnology. They fail to make reference to the many specific applications in fields such as agriculture, food processing and aquaculture. If there has been a specific focus, it has tended to be in the area of human genetic engineering. Studies dealing with public acceptance of science have shown that while the population may be clearly favourable to science, there "can be specific dissent about specific questions" (Yoxen and Green, 1989). Broad questions about acceptance of biotechnology may not provide us with many clues regarding how people think about some very specific applications. The above, combined with the fact that the term "biotechnology" encompasses a wide variety of applications and products and at the same time is defined in a number of different ways by scientists, make it problematic to assess the state of public opinion on biotechnologies as a whole; it also makes it questionable whether any such general attitudes to biotechnology are meaningful.

Public opinion surveys

Recent social science survey research shows that the public is not well informed and lacks detailed knowledge about new biotechnologies and their specific applications. This conclusion represents an oversimplification of a much more complex situation and may partly be due to the reasons mentioned above; confusion in terms of definitions and the fact that in some of the surveys the questions posed are too general. There are now a number of major quantitative studies on attitudes and acceptance of biotechnology available and undertaken in various western countries - reference will be made here to the surveys carried out in Denmark (Borre 1990), Ireland (Landsdowne 1989), Spain (IESA 1990), the USA (OTA 1987) and Europe (Càntley 1987, Eurobarometer 1991). Perhaps not surprisingly these quantitative surveys of public opinion contradict one another and do not give us a picture of common attitudes to genetic engineering - this is partly for methodological reasons (the surveys ask different questions) and also because of the political and cultural context in which they were undertaken. It can be argued that "public opinion" is a shallow concept and that in relation to "esoteric" subjects like genetic engineering where the public is generally not knowledgeable, opinions are formed at the time the questions are asked - converting them almost into "non-opinions", in that they had not been thought about, or held deeply. While polling techniques may reflect opinions about a given topic at a specific time, they have virtually no predictive value in terms of specific behaviour or action except perhaps for voting behaviour (where the questions asked are much more concrete and of considerably less conceptual complexity).

Having made the above qualifications, a broad finding common to all the studies is that the public tend to identify biotechnology today with genetic engineering and reproductive technologies used in human beings. Other areas such as deep fermentation or brewing are much less significant for the public.

The American study on "Public Perceptions of Biotechnology" undertaken by the Office of Technology Assessment (1987), which involved a very lengthy telephone interview questionnaire, revealed that most respondents expressed positive attitudes to science and technology in general. However, attitudes to biotechnology indicated a certain ambivalence.

Sixty six per cent felt that genetic engineering would improve life (compared with 92% for solar energy, 51% for nuclear power). At the same time 42% felt that it was "morally wrong" to change the genetic make-up of human cells, while only 24% questioned genetic engineering in relation to hybrid plants and animals. Table 1 shows that there is a continuum from acceptable manipulation when it comes to plants and bacteria to unacceptable when human cells are involved. When specific applications are mentioned this continuum is often inverted as is demonstrated in the Spanish study (IESA 1990). Respondents were prepared to accept ecological risks from the release of genetically modified organisms, however opposition was higher if such releases were in the local community and reflected the sentiment of "Not in My Back Yard" (Tait, 1988). The needs for regulation and protection from certain applications of biotechnology which might have negative environmental and health impacts were also on the agenda.

TABLE 1

"ON A SCALE 1 TO 10 WHERE 1 IS TOTALLY UNACCEPTABLE AND 10 IS TOTALLY ACCEPTABLE, WHERE WOULD YOU RANK GENETIC MANIPULATION OF...

Average acceptability of genetic manipulation of:

	Human cells	Animal cells	Bact. cells	Plant cells
	4.5	5.3	5.6	6.6
Science understanding:				
Very Good	5.2	6.1	5.9	7.2
Adequate	4.5	5.3	5.6	6.6
Poor	4.1	4.9	5.4	6.2
Heard about genetic engineering:				
A lot / Fair amount	4.9	5.9	6.0	7.2
Relatively little	4.3	5.2	5.4	6.3
Almost nothing	4.3	4.7	5.2	6.0
Effects of genetic engineering:				
Better	5.1	5.8	6.1	6.8
Worse	2.9	4.1	4.3	5.9
Religious:				
Very	4.4	5.2	5.5	6.3
Somewhat	4.5	5.3	5.7	6.8
Not too / Not at all	5.1	5.9	5.8	7.2

Source : (OTA, 1987)

A comparative study undertaken by the European Economic Community as early as 1979, indicated that there were significantly different attitudes towards genetic engineering according to country with 49% in Italy stating that genetic research was "worthwhile" compared with only 13% of Danish respondents (see table 2). Evidence of diversity was further reinforced by the Eurobarometer survey of 1991 (Table 3).

TABLE 2

COUNTRY	EC	B	DK	D	F	IR	I	L	N	UK
QUESTION:										
Genetic research:										
Worthwhile	33	38	13	22	29	41	49	37	36	32
Of no particular interest	19	20	10	16	22	20	19	31	17	21
Unacceptable risks	35	22	61	45	37	22	22	18	41	36
Don't know	13	20	16	17	12	17	10	14	6	11

Source : Cantley, 1987

TABLE 3

QUESTION: Science and technology change the way we live. Do you think that biotechnology and genetic engineering will improve our way of life in the next 20 years, it will have no effect, or it will make things worse?

COUNTRY	B	DK	D	GR	SP	F	IR	I	L	N	P	UK	EC
Will improve	49.2	43.6	43.7	39.1	57.9	53.7	48.4	55.5	48.3	48.3	46.4	51.2	50.4
No effect	8.1	9.3	18.6	2.5	3.3	10.1	10.2	5.8	13.7	7.9	2.4	7.0	9.6
Will make things worse	12.1	23.5	12.2	5.9	4.3	12.9	5.0	9.9	13.7	19.4	2.0	13.5	11.2
DK	30.3	23.6	24.6	52.5	34.5	23.3	36.3	28.8	24.3	24.4	49.0	27.6	28.4
No answer	0.4	0.1	0.9	0.0	0.0	0.0	0.0	0.0	0.0	0.0	0.3	0.8	0.4

Source : INRA Europe 1991

A Danish study (three separate surveys) carried out between September 1987 and May 1989 in the context of a major parliamentary debate was based on the OTA questionnaire, but in this case more focused on "gene technology". Ambivalence, at the aggregate level, towards genetic engineering is perhaps the highlight of this Danish survey: "A large majority believe that it is wrong to interfere with the genetic structure of higher organisms. However, an almost equally large majority feel that questions on gene splicing are best left to the experts... A somewhat smaller majority believe that it is important not to get behind in technological development. But this is countered by a majority who indicate that they would protest against work on gene technology being conducted in their neighbourhood. The proposal of an international ban on gene technology receives support from a sizeable minority." (Borre 1990: 5).

An Irish study (Landsdowne 1989) undertaken in 1989, using a much simpler and shorter questionnaire than the Danish or American surveys indicated a very tenuous public awareness of developments in the new biotechnologies. In response to an open question, most of the respondents were not able to state a single application of biotechnology. This obviously contradicts many of the findings of the Danish research (there is clearly a problem of comparability, given that the questionnaires were different as was the context in which the studies were carried out).

In the case of the Spanish study (IESA 1990), carried out by telephone in 1990, the most notable element is the concern expressed about genetic engineering in humans (49% finding genetic engineering unacceptable in human cells and 64% in human embryos). Views about genetic engineering in relation to plants, animals and bacteria were positive (the proportions saying it was acceptable:81%, 61% and 78% respectively). Women were in general more opposed to genetic engineering on all counts compared with men (15 percentage point difference). Practising Catholics were significantly more opposed to genetic engineering than were agnostics. The level of educational attainment represented another relevant variable: less education was linked with greater hostility towards biotechnology in general and genetic engineering in human cells in particular. The highly educated seem to have a more positive predisposition to biotechnology at an aggregate level - these data can be misinterpreted or misleading precisely because it is also from the more educated section of the population that we find the most organised and articulate opposition to biotechnology in the form of environmental pressure groups.

In this study in Spain, an interesting contradiction arose about genetic engineering in plants and animals; 76% and 72% were against the use of biotechnology for the development of more productive cattle and larger fish, yet the overwhelming majority stated that they were not concerned about genetic engineering in plants and animals. Respondents favoured a general line of scientific research (genetic engineering in animals) but were not happy about the concrete applications geared to the market (genetically engineered growth in animals). Positive attitudes were found to general questions about biotechnology (74% stating that it is beneficial) although negative feelings were expressed about some concrete applications. Furthermore, many respondents were unsure about what constitutes biotechnology: there has been some public debate in Spain, but the mass media has concentrated heavily on informing the public about developments in reproductive technologies (e.g. choosing the sex of a child and its legal consequences) rather than on agriculture, where the applications in Spain could be significant.

While not an opinion poll (although using a large sample) a recent study undertaken in the Netherlands by the SWOKA Institute for Consumer Research has tried to measure consumer acceptance of genetic engineering in foodstuffs using a system of consumer panels (870 respondents) and telephone-interviewing. The research indicated that attitudes were strongly related to the information provided. Information about the production method or hypothetical risk of biotechnologically-made foodstuffs resulted in a lower acceptance rate of the food product involved. In their concluding remarks the authors of the SWOKA report state:

"The public at large is not yet convinced of the need and merits of new biotechnology. This somewhat suspicious attitude can be relaxed when the benefits become clear and when truthful information (including well defined risk analysis) becomes available.... In the present situation we can not yet predict the outcome of the acceptance process; very much depends on a wise and responsible implementation of the technology, taking concerns and needs of the large public of consumers into account" (Hamstra 1991).

Most of the opinion poll studies have asked questions about sources of information on biotechnology as well as about who should regulate. In a recent Eurobarometer survey (1991), when asked to state in which sources of information they had confidence to tell the truth about biotechnology and genetic engineering, it was the consumer and environmental organisations which received the highest score with industry being placed very low. Significantly, in the Spanish survey industry was seen as being the least fit organisation/institution to regulate activities in genetic engineering.

Other quantitative studies have been undertaken in a number of European countries under the auspices of biotechnology enterprises; however the data have, in general, not been published or made available to the public.

Qualitative Approaches

Clearly the socioeconomic, cultural and political settings of the above mentioned countries are very different, and it can be argued that divergent results are a product of markedly different societies, values and systems of access to information. Another conclusion of an analysis of the quantitative studies of public opinion might be that the methodology provides an incomplete, even misleading picture of public attitudes and acceptance of biotechnology. Other (qualitative) approaches can be used, such as discussion/focus groups, workshops and in-depth interviews, to sharpen the picture. The quantitative studies have provided some tantalizing clues and have shown the ambivalence of public opinion to new biotechnology and its applications. The data which they have provided have been used as the basis of the design of much of the most recent qualitative research and vice versa.

It was precisely because of the contradictory and somewhat confusing results of earlier survey research that The European Foundation for the Improvement of Living and Working Conditions and Directorate General XI of the European Commission opted to support some qualitative research to gain a better understanding of the complex set of attitudes to biotechnology identified in earlier quantitative studies. This has involved research in Britain, France, Germany and Spain and was based on focus/discussion groups with members of the "informed public" in each of the four countries as well as workshops involving representatives of pressure groups and interested parties (pharmaceutical and food processing industries, agricultural interests, medical profession, trade unions, environmental groups, public administrations).

Focus/Discussion Groups

In each of the above mentioned countries two focus groups (one of men and the other of women) with 7 to 9 participants were held. The focus group technique is widely used by market research companies to assess public reaction to new products. It involves bringing together participants who are asked to give their opinion freely on the new product or theme. The discussion is led by a psychologist, social psychologist or sociologist trained in group dynamics and who acts as the group's moderator and follows a carefully sequenced discussion guide. Initially, no material about specific biotechnology products were presented by the moderator, instead the participants were asked to develop their own definitions of biotechnology and genetic engineering. Later on in the discussion, three products (out of biopesticides, herbicide resistant plants, genetically modified fish, BST and DNA probes) were presented and debated. Participants in the groups were chosen on the basis of a recruitment questionnaire. The purpose of the recruitment was to bring together members of the "informed public" - that is to say people with some university training, wide general knowledge and interest in science and technology, but not professionals in the field, politicians or members of pressure groups with an explicit interest in biotechnology. It should be clear that these "informed" participants represent only one segment of the "general" public. Generalisations about public perceptions from such studies are neither possible nor intended, although these groups may provide the analyst with interesting clues about some underlying and latent concerns.

The Sociopolitical Context

Before presenting the findings of these workshops it is necessary to make some preliminary remarks on the context in which the focus groups were conducted. In the United Kingdom, the groups were held at a time of three major food-related scares: **Salmonella** in eggs, **listeria** in some milk products and "mad cow disease". This had engendered a general feeling of vulnerability among consumers and growing distrust in the capacity of governments to establish acceptable controls and regulation in public health. At the same time legislation was being discussed in Parliament about experimentation on human embryos. In Germany there has been an important "greening" of the political discourse which includes all the major parties and not only the Greens. Germans read in the press nearly every day about research on genetic diseases and genetic engineering in the areas of medicine and agriculture.

In France and Spain, the theme of genetic engineering has less visibility than in the other two countries although the subject is now treated and debated more frequently in the press and other media. In Spain there have been major scandals related to the adulteration of food (cooking oil and illegal hormones in cattle).

Knowledge of Biotechnology

In all four countries, participants' awareness of biotechnology was clearest in relation to genetic engineering in the medical, reproductive and health care areas. Much less knowledge was evident with regard to agricultural and food biotechnologies. The exception to this was the German groups which seemed to have a clearer understanding of the role of biotechnology in these fields of production.

Most participants appeared to understand the concepts of genetic engineering and genetically modified organisms. Women tended to refer more than men to genetic engineering applied to humans.

In Germany awareness of new agricultural developments due to genetic engineering was quite high: new plants and animals, species made more resistant to environmental pressures, modification of the processes of growth and maturity in plants and animals were mentioned. Most German respondents were familiar with new developments in the field of food manufacturing. New medical developments due to genetic engineering were what participants knew most about: vaccines; insulin; possible cures for cancer; research on chromosomes and heredity in general; and determination of sex were all mentioned. Waste processing with the help of genetic engineering was the subject that participants knew least about.

In the British case the men mentioned a number of developments related to genetic engineering in different fields such as agriculture, food processing and health care. Women, on the other hand seemed to be less informed about developments, except in relation to health care. However, there was some question, in both groups, as to which developments were actually due to genetic engineering and which were due to other, more traditional, technologies.

The French and Spanish participants knew a lot more about developments in the medical field than in agriculture, food processing and the treatment of waste. **In vitro** fertilization, diagnosis of disease, cancer and AIDS research, hormones, cloning and vaccines were discussed as areas in which genetic engineering was making a major contribution.

Assessment of Benefits

New developments and applications in biotechnology were debated in various fields, of which some examples were discussed in more detail. The assessment of benefits was very similar for the four countries and many new biotechnologies were seen in general as improvements with the following **positive** outcomes for society:

*   Increase in food production, with particular relevance to the Third World, would result in a significantly higher standard of living.

*   Advances in detection of genetic diseases and introduction of new therapeutic techniques, the outcome of which would be better health.

*   New techniques to control and treat waste (especially toxic substances) were seen as a possible environmental benefit.

Economic benefits were mentioned on several occasions but there was nearly always someone to express reservations about the real motives behind these benefits. Is it for the good of humankind or exclusively for financial gain and profit making? Some participants expressed doubts that underdeveloped countries, which most needed the supposed benefits of genetic engineering, would be the ultimate beneficiaries.

Assessment of Risks and Safety

All the groups discussed risks at length. Strong expressions of concern were centred on the balance of nature (ecological equilibrium) and whether certain biotechnological products were environmentally friendly (Levidow and Tait 1990). It was repeatedly stated that there is insufficient knowledge about the long term effects of the release of genetically modified organisms on the environment. Other doubts were raised about the following:

*   The groups felt that research could get out of hand and that adequate institutional controls were lacking.

*   Certain products, such as herbicide resistant plants, with apparent economic and production benefits may have negative and unpredictable health outcomes.

*   BST in particular was not seen as necessary, because of existing surpluses and because of possible negative health outcomes for treated cows given the hormone as well as for the consumers of their products.

*   Participants in the Spanish groups expressed their unease about possible negative health and safety outcomes resulting from the use and consumption of genetically engineered products. There was also apprehension about the possible increase in inequalities (high status groups would receive benefits of biotechnology but not others who couldn't afford it). The modification of certain social, family and institutional roles as a result of the application of biotechnologies related to human reproduction was considered to represent a risk.

* In the two German groups the list of perceived concerns was longer than the list of benefits and showed that participants were worried about the possible consequences of all the new applications. The main concern was certainly the lack of control and the unpredictability of long term outcomes. In many cases the respondents could not decide whether the evolution of science in general and genetic engineering in particular was a good or a bad thing.

* The French participants were, altogether, the most positively inclined towards biotechnology. Concerns and risks were linked with questions about the consequences of research in the long run, such as the disappearance of some species, imbalance of the ecological system, and the increasing gap between developed and underdeveloped areas of the world.

* The concerns and risks discussed in the two British groups were very general and included the long term effects on the ecological system, the reliability and competence of those in charge of research, the lack of control over the evolution of these matters and their economic implications for business, consumers and the Third World.

Access to Information

Information access was one of the themes highlighted in the focus groups. There was a surprising degree of common opinion despite the fact that the groups were held in four countries with rather different traditions in the information field. Information about new biotechnologies seems to be a priority at least for the "informed public".

* In all the groups the participants underlined the general lack of accurate, understandable and easily accessible information on subjects related to genetic engineering and they felt that they would like to know more about the research, its outcomes and more particularly its possible economic, social, ethical and environmental implications for their lives.

*   In what way will genetic engineering effect everyday life?

*   What is the state of consumer acceptance and how will it evolve in relation to emerging applications?

*   What are the most appropriate decision-making structures regarding research development and use of genetic engineering.

*   How can developments in genetic engineering and potential applications best be rendered intelligible to the public at large?

In all four countries difficulties arose both before and during the workshops. Some of the convened participants expressed their reluctance to attend, lacking confidence in the purpose and motives behind the organisation of the workshops. Was it a "marketing" operation to sell biotechnology or to gain knowledge of critical positions in order to provide more acceptable information on biotechnology? Would the non-scientists be competent to discuss in depth the technical issues related to safety? Would the meeting not become "politicised"? Would the moderator be impartial? Such misgivings represent a major handicap because it is a condition for the success of such a workshop that those taking part should have confidence in the impartiality and objectivity of the organisers in order to be able to contribute to the discussion in a frank and open way. In the course of the workshops not all these doubts were dispelled, making it difficult to achieve dialogue. In general the discussions were less fluid than the "focus" groups with confrontation and tension rather than dialogue and interaction as a background. Nevertheless, they revealed the divergent discourses and some of the underlying tensions between the major players in biotechnology and in this sense the workshops proved to be of considerable interest.

Context

In the four countries, the state of regulation and political debate on biotechnology varies considerably. The German groups were held shortly after the passing of the Genetic Engineering Act (28th March 1990). At the same time, environmental issues have held a central place in food supply, there seems to be some lack of public confidence in food processing and agricultural biotechnologies in the context of recent but unrelated issues such as BSE and salmonella in Britain and illegal growth hormone use for cattle fattening in Spain.

The Workshops

The objective of the workshops was to develop an exchange of information and opinion between parties involved with genetic engineering and interest/pressure groups. It was intended that this should ascertain the current state of the debate in the four countries and possibly facilitate exchange of priorities and concerns. In the process, the ways and means by which the social dialogue between the individual interest groups could be extended and intensified were examined. Therefore, representatives of science, industry, medicine, public administrations, consumer associations, environmental groups, feminist organisations, trade unions, churches and other social groupings were invited to attend the workshops. In this way it was intended that the participation should reflect as wide a spectrum of opinion as possible.

Organisational and Technical Problems

The aim was to create discussions which would be demanding in terms of subject matter and content. The dialogue would be between highly informed and articulate participants accustomed to debating issues around biotechnology with "friend or foe". The groups were in general kept small (with the exception of the British workshop) to allow for greater individual participation. Many of those invited to the meetings held senior positions in their organisation and had represented them in public fora on biotechnology.

In principle, a brief exchange of information on the latest developments in research and commercial applications of genetic engineering was to take place followed by discussions of some polemical issues such as ethics, social impact, risk assessment and so on. The moderator was to place particular emphasis on consumer-oriented questions, such as:

Attitudes and Diversity: Gender and Country Differences

Not surprisingly, differences were found between the groups according to gender and country. With regards to the former differences, in most cases women were more critical and sceptical than men about biotechnology, especially in relation to human genetic engineering. However, the women's groups appeared to trust scientists more where medical and health care was concerned compared with men. This applied not only to issues related to biotechnologies in the area of reproduction but also to the food processing industries and agriculture with very explicit concerns being raised on possible safety and public health outcomes. An exception to this pattern of gender differences was found in Spain, where there seemed to be a concern about the impact of human genetic engineering on social and family roles and dynamics amongst men more strongly than among women.

In relation to country differences, in Britain and Germany greatest interest was focused on ethical and medical implications of genetic engineering. However, the environmental risks of biotechnology applications were clearly articulated (especially in the German groups) as were concerns about the lack of regulatory measures and implementation of existing, if apparently inadequate, legislative provisions. In the British case events in fields not related to biotechnology, but to other areas of food production, seem to have created great anxiety amongst consumers vis-à-vis the will or capacity of public administrations to protect public health and safety around agro industry and food production. In France, a rather positive evaluation of biotechnology was evident (this being paralleled in other areas of science and technology, the best known being on nuclear power), while in Spain the central concern was related to genetic manipulation in humans and particularly in the field of reproductive technology. The perceived lack of regulatory measures and state controls were very much on the agenda in the latter case.

Broadly speaking it is possible to talk of positive acceptance of the therapeutic applications of biotechnology (e.g. insulin) but with concern over possible ethical issues in medical research. At the same time, while there may be positive responses to the role of modern agricultural biotechnology in solving some of the pressing demands of the Third World

*   There should be a "right to information" on biotechnological research and commercial applications. It should be of easy access.

*   Information should, in the best of all possible worlds be objective, autonomous and free from the interference of industry or economic pressure groups (the source of information should always be made clear).

*   Some respondents felt unable to comment with confidence on biotechnology precisely because they felt that they did not have enough reliable and objective information. Of course the level, range and diversity of information present problems for meaningful and acceptable communication.

Regulation and Control

Regulation was another of the major themes raised in all the groups. There was an evident "lack of faith" in the competence and will of governments to control developments in genetic engineering or other areas of research where there are possible safety and public health risks. A feeling of vulnerability was manifest in all the groups - being quite independent of the ideological positions of the respondents.

*   It was repeatedly argued that regulation should be based on sound, honest and objective information about the usefulness of biotechnology and the possible risks attached to some of its applications.

*   Regulation should be free from the interference of commercial or industrial interests.

*   Some groups proposed that supra-national organisations play a more active role in the regulation and control of new biotechnologies.

political debate now for over a decade and have influenced the discourses of all the major political parties, consumer associations and trade unions.

In Britain, apart from the BSE and **salmonella** issues already mentioned, specialists and interest groups have also been debating the use of BST and the setting up of a Food and Drugs Commission independent of the Ministry of Agriculture. The proposal to set up a Council for Biotechnological Safety similar to the Council for Nuclear Safety has been the most important legislative initiative in Spain. The rather more pro science and technology atmosphere in France is reflected in attitudes (public and interest groups) to genetic engineering.

Issues and the Debate

While an attempt was made to keep to the same format for discussion in the four countries, the debate developed rather differently in each case, although there was a common core of issues which were raised in all the workshops. These included (not necessarily in the order given below):

*   Information

*   Decision-making, regulation and control

*   Public participation

*   Ethics

*   Socioeconomic impacts

*   Health and safety in the workplace

*   Environmental considerations

Information

We have already seen that the information issue was the "star" theme in the "focus" groups amongst the interested public. The workshops were no exception and many of the interventions of articulate professionals, experts in the field of biotechnology, paralleled those of the non-specialist public. It was understood by all the participants that the general public was eager for information on biotechnology but with certain conditions attached:

* Information should be provided in clear, accessible and understandable language, and not couched in lofty specialist terminology.

* It should be objective and truthful.

* It should come from a reliable source which should be identified.

The above were seen as prerequisites for strengthening public confidence. This is as far as consensus went. Representatives from industry tended to argue that the facts spoke for themselves. Information could not always be provided because of the need for confidentiality in the face of competition in the development of new products for the market. Interest groups in this context often gave the impression that industry regarded the principle of secrecy as more important than the provision of information. Consumer organisations and environmental groups were highly critical of current information policy in biotechnology which they considered to be inadequate. Access to information was a right and there should be mechanisms for guaranteeing this where risks for the environment, health and safety may be involved. The withholding of information would only reinforce the existing lack of confidence in commercial activities in the scientific and technological field.

The role of the press in informing the public was discussed in the Spanish groups. The extent to which the information was objective or biased became the main issue of debate. Both industry and consumers claimed that the press often had an "axe to grind" and that the level and quality of information was sometimes inadequate. Journalists argued that they were unjustly criticised and even persecuted by interest groups. Nevertheless it was recognised that the mass media should have a role in disseminating information and debate on biotechnology and its commercial applications - it was not clear from the discussion precisely what this role could be.

Decision-Making, Regulation and Control

This was another area of debate which soon became polarised in most of the workshops. Environmental groups underlined that they and the public at large felt unprotected and that the legal framework for genetic engineering was inadequate. Even the Genetic Engineering Act (in Germany) left many gaps, both with regard to the approval of genetic engineering processes and the marketing of genetically engineered products.

An intermediate position, yet still critical, argued that appropriate decision-making structures could help in reducing public uncertainty and anxiety about developments in biotechnology. At present no consideration in the regulations was given to socioeconomic or ethical questions. There were as yet no seriously developed criteria governing the conduct of cost-benefit analyses.

From industry it was argued that when it comes to regulating biotechnology, the focus should be on the product, which should be judged according to the already established criteria of quality. At the same time over-regulating could hinder research and development as well as representing a barrier to competition.

Public Participation

Consumer groups were insistent that the public should be involved at an early stage in the decision-making process. In most cases, they argued, they were not represented on committees which have a mandate to decide on products or processes. As a result the public had to rely on the efficiency and objectivity (placed in doubt) of public administrations. Both the questions of information and regulations, it was argued, had to involve the public and consumer organisations.

Doubt was cast on the objectivity of many public interest organisations (consumers, environmentalists) and therefore their ability to participate seriously in decisions on the future of biotechnology and particularly the marketing of genetically engineered products. For some representatives of industry and science some of the consumer and environmental groups had a hidden political agenda making interaction between these pressure groups, industry and public administrations difficult. They went as far as to say (in the German case) that there was often deliberate manipulation of public opinion to turn it against industry. In such cases, genetic engineering often served simply as an instrument for other quite distinct political objectives. This industry view of the manipulative role of certain groups was seen as an attempt to justify the cutting off of dialogue and cooperation between the public and industry. It was stated that these negative attitudes among those involved in profit-making ventures reflected a lack of interest in establishing dialogue, since this could threaten their commercial expansion. This discussion in the German groups and to a slightly lesser extent in the Spanish workshops, was another point of maximum polarisation between representatives of industry and consumer and environmental groups.

Ethical Issues

The main ethical issues centred around human genetics. The questions raised by the consumer and environmental organisations and representatives of religious groups were very similar to those of the informed public in the "focus" groups. Apprehension and anxiety were expressed about the manipulation of human genetic material even when diagnostic benefits could be demonstrated. While the therapeutic and diagnostic applications found much support there was concern about the use of genetic information; social pressure to have an abortion in the face of negative prenatal diagnostic information was an example cited. It was pointed out that this was a problem not necessarily strictly related to genetic engineering, since prenatal information could be gained with the use of other technologies and that the public often confused these other techniques in human reproductive technology with genetic engineering. Concern was expressed about genetic screening at work in relation to the right to privacy.

Environmental Considerations

The environment was recognised by practically all the participants as a concern of the public. All the public interest sectors argued that there were major risks involved in the release of genetically modified organisms. Deliberate or accidental release could lead to alterations in the delicate balance of nature, and could cause irreversible damage. Even the provision of formal controls and regulations governing release projects could not conceal the fact that a reliable system of risk assessment is not possible. Despite the above remarks there was also criticism of existing control measures and a demand for new statutory provisions based on a comprehensive environmental risk management policy as well as on effective controls; this possibilistic position contradicts the notion that there could be no such thing as a "fool proof" system of risk assessment. The question of liability in the event of damage would also need to be resolved. The analogy of the "sorcerer's apprentice" was used to underline the risks of trying to "play" with technology - things can all too easily get out of hand ending up in disaster according to the most critical groups. It was pointed out by some participants that genetic engineering could be used in the field of environmental improvement and protection.

Socioeconomic Impacts

Many of the benefits of modified organisms were overlooked according to those in favour of the promotion of genetic engineering. It was especially relevant to consider the contribution to solving the world food supply and the desperate situation in the Third World. Acrimonious debates ensued as to who really benefits from biotechnology with some participants casting doubt on the ability and particularly the will of industry to contribute to the resolution of such problems. Some felt that biotechnology could widen the gap between rich and poor.

Many other socioeconomic implications were mentioned but not developed in depth. They included such polemical issues as: patenting of modified organisms; the future of farm structure - would the small farmer survive; the manufacture of biological weapons (this issue was raised also as an ethical question). It was clear for the consumer groups that many of the socioeconomic effects had an extremely strong ethical component: genetic fingerprinting; genome analysis; gene therapy.

In this field, as in so many, views were often polarised and dialogue would not be an appropriate description of the interaction which took place.

Health and Safety in the Workplace

While this subject has frequently been the basis of vigorous debate between industry and trade unions, little time was devoted to the subject in the country workshops; although there was a general consensus that safety at the workplace was of prime importance in relation to genetic engineering. Appropriate safety measures and instructions would have to be established both in the laboratory as well as in manufacturing processes. The necessary extent of government intervention and control of workplace safety was an issue that separated participants, for example trade unions and business.

Comparing and Contrasting the Focus Groups and the Workshops

In the more confrontational workshops there was greater similarity in the outcome of the debates than was the case with the "focus" groups. In the former, positions were largely predetermined, inflexible and structured (perhaps this should not have come as a surprise and should not be interpreted as a failure of the workshop technique). In this context few country differences emerged; the discourses of each interest group were remarkably homogeneous across countries, with little apparent underlying diversity. The French workshop may be an exception, given that the critical participants demonstrated greater affinity with advances in science and technology than in the other three countries. The manifest tensions contrasted strongly with the fluidity of discussions in the "focus" groups. Here it was clear that while there were many common concerns across countries there were also specific preoccupations for each country related to differing sociopolitical backgrounds as well as cultural diversity.

Although similar workshops have been organised around equally polemical issues in the past with some success (nuclear power for instance), in the case of genetic engineering the subject proved to be much more difficult to debate. It is possible that the use of in-depth interviewing of representatives of the main concerned parties would have provided a eomplimentary picture of contrasting positions, though the extent of polarisation may not have been revealed.

The Context and Implications of Concern

Public attitudes to genetic engineering are both complex and ambivalent. The economic potential of biotechnology is said to be considerable and it is also argued that it could contribute significantly towards resolving important problems in diagnostics, therapeutics, the food supply and environmental degradation. Despite public support for science in general and for scientific applications which imply major benefits for society, many environmental, socioeconomic and ethical concerns have been raised about some of the newer technologies and especially genetic engineering.

The public has expressed its apprehension about the impacts of biotechnology applications on the one hand and administrative control and regulation of genetic engineering on the other. Impact (socioeconomic and environmental), regulation and information appear to be three key words, all of them related to a growing perception of vulnerability in the face of technological change and environmental insecurity.

The evolving perception of insecurity, especially amongst the "informed" public is related, at least in part, to the extension of the environmental discourse which in its most radical form "presents nature as a fragile, vulnerable system, constantly under threat from chemical agents out of control" (Levidow and Tait 1991: 15). Changes in the public's perception of the environment and the growing sense of vulnerability is not just a fashion (Lemkow and Buttel 1983): during the early sixties, some commentators were painting a bleak picture of the future planetary environment. Rachel Carson in her best-selling **Silent Spring** was telling her largely middle class readership that:

"For the first time in the history of the World, every human being is now subjected to contact with dangerous chemicals, from the moment of conception until death. In less than two decades of their use, the synthetic pesticides have been so thoroughly distributed throughout the animate and inanimate world that they occur almost everywhere... they have been found in fish in remote mountain lakes, in earthworms burrowing in the soil, in the eggs of birds - and in man himself. For these chemicals are now stored in the bodies of the vast majority of human beings regardless of age. They occur in the mother's milk and probably in the tissues of the unborn child" (Carson 1963: 15-16).

These pollutants (synthetic pesticides, organic mercurial and other compounds of heavy metals), the product of new techniques, were not visible in the dramatic form of industrial smog, but appeared to be more insidious with greater risks for the environment and human health. The new products, because of their mobility, were capable of affecting all social classes including those which had come out relatively unscathed from the worst effects of industrialisation. A growing section of these new groups, fundamentally of high socioeconomic and educational status, began to perceive these changes as a threat - to feel vulnerable and insecure in the face of qualitative changes taking place in the environment, many of them induced by the use of new products. It is in this context that genetic engineering has been placed today by sections of the public and is therefore viewed as a possible agent of environmental insecurity as well as having some negative implications for public health. It is not surprising then, to discover that the anxiety produced by genetic engineering has fostered new strategies on behalf of firms involved in biotechnology. This includes selling their products as "environmentally friendly" and projecting new biotechnologies as "evolutionary" rather than "revolutionary" developments as part of what humankind has been doing for millennia: selective plant and animal breeding programmes; brewing beer and making cheese and yogurt (Levidow and Tait 1991).

While "tampering with nature" is an environmental concern, it also holds a strong ethical component in relation to genetic engineering in humans. Social control and anti-utopian ("dystopian") visions of the future can be found where the Huxlian universe of **Brave New World** (1932) seems to be within the reach of gene technology. Less pessimistic yet ambivalent visions of futures dominated by genetic engineering have been addressed by a new generation of mainstream and science fiction writers dealing mainly with the social implications of this technology (Bujold 1988, Butler 1987, Cherryh 1988, Herbert 1987).

Moving on to another key concept, regulation, it is clear that the regulatory climate of the 1980s and 90s in many countries contrasts notably with that of earlier post war decades. Republican and Conservative administrations in the USA and Britain have ushered in a return to philosophies of economic liberalism where deregulation and flexibilisation are considered to hold the key to economic growth and "social progress": more market forces, less bureaucratic regulation and greater consumer choice. Today, such approaches to regulation appear to be increasingly acceptable even in countries where there has been a tough tradition of regulation and control.

In the current deregulatory atmosphere there is growing evidence of deepening public concern about the lack of control of economic activities which have implications for the maintenance of environmental equilibrium and public health. The European Foundation/DG XI study has indicated that this preoccupation is independent of ideological positions so that conservatives, liberals and progressive alike, while having differing attitudes to regulation in general, all seem to argue for tighter controls in what they consider to be two vital areas: public health and environment. At the same time there are signs that in these fields the public questions the will and even the ability of administrations to control activities which imply environmental and health risks. As was pointed out earlier, the experience of "mad cow disease" (Britain) and the use of illegal growth hormones in cattle (Spain), although formally not linked to genetic engineering have undermined public confidence in regulatory mechanisms. It should be added that the root of this waning of public confidence is not exclusively related to the objective regulatory efficiency of public administrations. Not only does it appear to be linked to changing perceptions of the impact of technological change, but also, as already insinuated, to the "greening" of political discourses (Yoxen and Green 1989) in the context of new environmental risks some of which can be caused by technological innovation.

Closely related to regulation and control is the issue of information. This does not necessarily mean making larger numbers of better and more understandable audiovisual materials, which is often the way the question is addressed. The deliberate release of genetically modified organisms or the introduction of a new genetically engineered product again raises the problem of public confidence and the demand for access to appropriate information. One group of scientists has referred to the problem as "coping with the scarcity of information". They go on to state:

"There will be many decisions where the scientific component is far from conclusive, and so the skill and integrity of those making the decisions, as perceived by the public, will be crucial. It is for this reason that we advocate as much clarity and publicity as possible in the management of uncertainty in the control of the hazards of genetically engineered microorganisms" (Finchmann and Ravetz 1991).

Agenda for Discussion

Three main issues seemed to emerge from debate and discussion on genetic engineering at meetings held in the EC framework.

Future Research Needs

While opinion polls play a significant role in improving our understanding of public perceptions of biotechnology, they can at best only provide a superficial impression of the state of opinion on scientific applications in such areas as genetic engineering. There is a need to diversify research options by introducing some of the "qualitative" methods already outlined in this document (discussion groups etc.). At the same time social and cultural diversity should be taken account of when designing research in this field. The "public" is not a homogeneous entity; there is a marked diversity in perception according to the socioeconomic and other circumstances of the consumer in general and the principal actors in biotechnology in particular. Therefore more focused research on the concerns of particular groups (from, for example "green" consumers to industry and trade unions) should be undertaken. This should also include analysis of the perceptions of government and policy makers. Furthermore, given that public funds are frequently used for such studies, it should be made clear what the objectives of the research are. A suggested agenda for future analysis could include:

* Documentation of preferences and not only attitudes among different groups of the 'public'.

* Clarification of the relationship between the so-called "underlying" attitudes to science in general and perceptions of biotechnology.

* The need to distinguish more clearly attitudes to different applications of biotechnology: food, therapeutics, agriculture, environment.

* Analysis of the (in)stability of public perception of biotechnology over time.

* The need to take into account specific concerns: ethical safety, health questions.

The research process should be transparent and open with it being clear what its objectives are and who has paid to do the research.

Information and Strategies for Participation

The issue of communicating science clearly to the public, free from the interference of interested parties remains an area of concern for the scientific community as well as for the public. In this context it is also evident that the public has a right to define what it wants to know about new biotechnologies.

It has been suggested that a distinction should be made as to the purpose of information:

*   Information to promote biotechnology (e.g. Public Relations).

*   "Neutral" information, to help consumers to make balanced choices

*   Information for enabling public participation in the decisions about biotechnology.

It is a generally agreed principle that information should be provided in clear, accessible and understandable language, and not couched in lofty specialist terminology. It should be truthful and it should come from a reliable source which should be identifiable. This laudable, but somewhat naïve aspiration for trustworthy information, so frequently articulated as a consensus arising out of meetings between the main interested parties in biotechnology, has so far not resulted in many clearly defined strategies indicating how this might be achieved. Such conditions are necessary for strengthening public confidence. The issue of uncertainty is frequently raised and whether and/or how such uncertainty when dealing with risk assessment of genetic engineering applications should be communicated to the public. Specific experiences such as the Danish "consensus conference", while not necessarily applicable to all Member States could serve as a relevant precedent for participatory approaches to public information. Diversity is emerging as a topic of discussion: the public is not homogeneous and different sections of society have different information demands as well as different ways of assessing information. Diversity, in terms of beliefs, values and culture should be considered when analysing the adequacy of public information.

Regulation and Public Policy:

Regulation and public policy is intimately related to information. If public confidence in the safety and efficacy of new products is to be achieved, then transparency in information and regulation are essential. This implies the need for an open system of regulation. Some specific experiences can be mentioned where there has been secrecy and a lack of transparency in relation to the release of genetically modified organisms. Where appropriate information is not given to the communities when releases take place, local residents soon begin to distrust the administration. It appears to be frequently the case that public administrations are not committed to the public understanding of science and its applications. Information is the basis for public confidence and when it is provided in an open way demands for information by the public become much less vociferous.

Regulation is not just about having formal frameworks for supervising and controlling genetic engineering. "Justice must be seen to be done", otherwise the public will mistrust regulations and regulators. In parts of Europe there is a considerable gap between law and practice: there are copious amounts of "harmonised" EEC legislation and regulations but in some cases virtually no implementation of them. This again promotes a lack of public confidence in regulatory mechanisms. Health and safety problems which have occurred in areas not strictly related to biotechnology (**salmonella**) have created an atmosphere of mistrust, where the administrations are seen to be ineffective and lacking the political will to properly implement regulations. One of the very concrete conclusions of the Eurobarometer Survey (1991) was that there were demands for biotechnology to be regulated.

It is clear from the studies presented here that public attitudes to genetic engineering are ambiguous and complex. Perceptions have been formed in a sometimes tense and difficult atmosphere. In this context there does seem to be a need for public fora which could bring together the plurality and diversity of views and interests, essential for the future clarification of the many issues raised by genetic engineering. Such fora could also help to build up greater trust and confidence amongst those involved in the debate surrounding biotechnology.

At the same time one of the major concerns for both researchers and policy makers must be how to introduce mechanisms which would "put the public back into policy". What instruments might be developed for public participation related to possible impacts and implications of genetic engineering? How should the public be involved in evaluating new biotechnology products? What criteria should be used for the siting of biotechnology plants? How can the public and the administration best interact over the environmental and ethical implications of genetic engineering? All these questions have to be addressed precisely because there is significant underlying public unease about genetic engineering but no clearly articulated processes for bringing together policy makers and the public on this issue.

BIBLIOGRAPHY

BORRE O. (1990), **Public opinion on Gene Technology in Denmark 1987-89,** Aarhus University, Aarhus.

BUD R. (1989), **Janus-faced Biotechnology: an Historical Perspective** in Tibtech, September (vol. 7), Elsevier (UK).

BUJOLD L.M. (1988), **Falling Free,** Baen Books, New York.

BULL A.T., HOLTON G. & LILLY M.D. (1982), **Biotechnology: International Trends and Perspectives,** OECD, Paris.

BUTLER O.(1987), **Xenogenisis,** Warner, New York.

CANTLEY M.F. (1987), **Democracy and Biotechnology: Popular attitudes, information, trust and the public interest** in Swiss Biotechnology Nr. 5, Zurich.

CARSON R. (1962), **Silent Spring,** Miflin, New York.

CHERRYH C.J. (1988), **Cyteen,** Warner, New York.

EUROBAROMETER 35.1 (1991), **Biotechnology,** INRA for the Commission of the European Communities.

FINCHMANN J.R.S. & RAVETZ J.R. (1991), **Genetically Engineered Organisms, Benefits and Risks,** Open University Press, Milton Keynes.

FLEISING A. (1989), **Risk and culture in biotechnology,** in Tibtech March 1989, Vol 7, Elsevier Science Publishers, (UK).

GOODFIELD J. (1977), **Playing God,** Hutchinson, London.

HACKING A.J. (1985), **Economic Aspects of Biotechnology,** Cambridge University Press, Cambridge.

HAMSTRA A. (1991), **Biotechnology in Foodstuffs: Towards a model of consumer acceptance,** SWOKA, Leiden.

HERBERT F. (1976), **The Eyes of Heisenberg,** Berkeley, New York.

HUTTON R. (1978), **Bio-Revolution: DNA and the Ethics of Man-Made Life,** New American Library, New York.

HUXLEY A. (1932), **Brave New World,** Chatto and Windus, London.

INSTITUTO DE ESTUDIOS SOCIALES AVANZADOS (1991), **Biotecnología y opinión pública en España,** CSIC, Madrid, (Informe).

JENKINS R., LEMKOW L., SNELL P. & SOTO P. (1991), **Elements for an Evaluation of Food Biotechnology from a Consumer Point of View: Final Report,** Consumer Policy Service of Commission of the European Communities.

KLOPPENBURG J.R. (1988), **First the Seed: the Political Economy of Plant Biotechnology,** Cambridge University Press, Cambridge.

KRIMSKY S. (1982), **Genetic Alchemy: A Social History of the Recombinant DNA Controversy,** MIT Press, Cambridge, Mass.

LANDSDOWN MARKET RESEARCH LTD. (1989), **Biotechnology - Awareness and Attitudes,** Dublin. (Report).

LAWRENCE W.W. (1985), **Modern Science and Human Values,** Oxford University Press, Oxford.

LEMKOW L. (1991), **Biotechnology and Public Opinion in Spain** in CSIC News, Consejo Superior de Investigaciones Científicas, Madrid.

LEMKOW L. (1991), **Comparative Study of the Current Provision for Public Consultation and Accessibility to Information in Biotechnology: Final Report,** DG XI of Commission of the European Communities.

LEMKOW L. and BUTTEL F. (1983), **Los movimientos ecologistas,** Mezquita, Madrid.

LEVIDOW L. (1991), **"Cleaning up on the Farm"**, Culture as Service, Vol. 2, Part 4, No. 13, FAB, London.

LEVIDOW L. & TAIT J. (1991), **"The greening of Biotechnology : GMOs as environment friendly products"**, Science and Public Policy, October

NATIONAL RESEARCH COUNCIL (1989), **Field Testing Genetically Modified Organisms: Framework for Decisions,** National Academic Press, Washington.

NELKIN D. (Ed. 1979), **Controversy: Politics of Technical Decisions,** Sage, London.

OECD (1988), **Biotechnology and the Changing Role of Government,** OECD, Paris.

OTA (1987), **New developments in Biotechnology: Public Perceptions of Biotechnology,** U.S. Government Printing Office, Washington DC.

RICHARDS J. (Ed. 1978), **Recombinant DNA: Science, Ethics and Politics,** Academic Press, New York/London.

ROBERTS E. (1989), **The Public and Biotechnology,** European Foundation for the Improvement of Living and Working Conditions, Dublin.

ROGERS M. (1977), **Biohazard,** Knopf, New York.

RUSSEL A.M. (1988), **The Biotechnology Revolution: An International Perspective,** Wheatsheaf Books, Brighton.

SARGEANT K. & EVANS C.G.T (1979), **Hazards Involving the Industrial Use of Microorganisms,** CEE, Brussels.

TAIT J. (1988), **NIMBY and NAIBY: Public Perceptions of Biotechnology** in International Industrial Biotechnology 8,6: (p. 5-9).

TEICH A.H., LEVIN M.A., & PACE J.H. (1985), **Biotechnology and the Environment: Risk and Regulations,** A.A.A.S., Washington DC.

TEITELMANN R. (1989), **Gene Dreams: Wall Street, Academia and the Rise of Biotechnology,** Basic Books, New York.

US GOVERNMENT (1984), **Commercial Biotechnology: An International Analysis,** US Government Printing Office, Washington DC.

WATSON J.D. & TOOZE J. (1981), **The DNA Story: A Documentary History of Gene Cloning,** W.H. Freeman, San Francisco.

WEBBER D.J. (1990), **Biotechnology: assessing social impacts and political implications,** Greenwood Press, New York.

WHEALE P. & McNALLY R. (EDS 1990), **The Biorevolution,** Pluto Press, London.

WOODHEAD A.D. & BARNHART B. (1988), **Biotechnology and the human genome: innovations and impact,** Plenum Press, New York.

YANCHINSKI S. (1985), **Setting Genes to Work: The Industrial Era of Biotechnology,** Penguin, Harmondsworth.

YOXEN E. (1983), **The gene business: who should control biotechnology?,** Harper Row, New York.

YOXEN E. & GREEN K. (1990), **Scenarios for biotechnology in Europe: a research agenda,** European Foundation for the Improvement of Living and Working Conditions, Dublin.

YOXEN E. and HYDE B (1987), **The Social Impact of Biotecnology,** European Foundation for the Improvement of Living and Working Conditions, Dublin.

YOXEN E. & DI MARTINO V. (Eds 1989), **Biotechnology in future society: scenarios and options for Europe,** Dartmouth (for the EEC), Aldershot.

European Foundation
for the Improvement of Living and Working Conditions

Public Attitudes to Genetic Engineering:
Some European perspectives

Luxembourg: Office of Official Publications of the European
            Communities

1993 — 54p. — 16 × 23.5 cm

ISBN 92-826-4002-7

Price (excluding VAT) in Luxembourg: ECU 6

Venta y suscripciones • Salg og abonnement • Verkauf und Abonnement • Πωλήσεις και συνδρομές • Sales and subscriptions • Vente et abonnements • Vendita e abbonamenti • Verkoop en abonnementen • Venda e assinaturas

BELGIQUE / BELGIË

**Moniteur belge /
Belgisch Staatsblad**
Rue de Louvain 42 / Leuvenseweg 42
B-1000 Bruxelles / B-1000 Brussel
Tél. (02) 512 00 26
Fax (02) 511 01 84

Autres distributeurs /
Overige verkooppunten

**Librairie européenne/
Europese boekhandel**
Rue de la Loi 244/Wetstraat 244
B-1040 Bruxelles / B-1040 Brussel
Tél. (02) 231 04 35
Fax (02) 735 08 60

Jean De Lannoy
Avenue du Roi 202 /Koningslaan 202
B-1060 Bruxelles / B-1060 Brussel
Tél. (02) 538 51 69
Télex 63220 UNBOOK B
Fax (02) 538 08 41

Document delivery:
Credoc
Rue de la Montagne 34 / Bergstraat 34
Bte 11 / Bus 11
B-1000 Bruxelles / B-1000 Brussel
Tél. (02) 511 69 41
Fax (02) 513 31 95

DANMARK

J. H. Schultz Information A/S
Herstedvang 10-12
DK-2620 Albertslund
Tlf. (45) 43 63 23 00
Fax (Sales) (45) 43 63 19 69
Fax (Management) (45) 43 63 19 49

DEUTSCHLAND

Bundesanzeiger Verlag
Breite Straße
Postfach 10 80 06
D-W-5000 Köln 1
Tel. (02 21) 20 29-0
Telex ANZEIGER BONN 8 882 595
Fax 2 02 92 78

GREECE/ΕΛΛΑΔΑ

G.C. Eleftheroudakis SA
International Bookstore
Nikis Street 4
GR-10563 Athens
Tel. (01) 322 63 23
Telex 219410 ELEF
Fax 323 98 21

ESPAÑA

Boletín Oficial del Estado
Trafalgar, 29
E-28071 Madrid
Tel. (91) 538 22 95
Fax (91) 538 23 49

Mundi-Prensa Libros, SA
Castelló, 37
E-28001 Madrid
Tel. (91) 431 33 99 (Libros)
     431 32 22 (Suscripciones)
     435 36 37 (Dirección)
Télex 49370-MPLI-E
Fax (91) 575 39 98

Sucursal:
Librería Internacional AEDOS
Consejo de Ciento, 391
E-08009 Barcelona
Tel. (93) 488 34 92
Fax (93) 487 76 59

**Llibreria de la Generalitat
de Catalunya**
Rambla dels Estudis, 118 (Palau Moja)
E-08002 Barcelona
Tel. (93) 302 68 35
     302 64 62
Fax (93) 302 12 99

FRANCE

**Journal officiel
Service des publications
des Communautés européennes**
26, rue Desaix
F-75727 Paris Cedex 15
Tél. (1) 40 58 75 00
Fax (1) 40 58 77 00

IRELAND

Government Supplies Agency
4-5 Harcourt Road
Dublin 2
Tel. (1) 61 31 11
Fax (1) 78 06 45

ITALIA

Licosa SpA
Via Duca di Calabria, 1/1
Casella postale 552
I-50125 Firenze
Tel. (055) 64 54 15
Fax 64 12 57
Telex 570466 LICOSA I

GRAND-DUCHÉ DE LUXEMBOURG

Messageries Paul Kraus
11, rue Christophe Plantin
L-2339 Luxembourg
Tél. 499 88 88
Télex 2515
Fax 499 88 84 44

NEDERLAND

SDU Overheidsinformatie
Externe Fondsen
Postbus 20014
2500 EA 's-Gravenhage
Tel. (070) 37 89 911
Fax (070) 34 75 778

PORTUGAL

Imprensa Nacional
Casa da Moeda, EP
Rua D. Francisco Manuel de Melo, 5
P-1092 Lisboa Codex
Tel. (01) 69 34 14

**Distribuidora de Livros
Bertrand, Ld.ª**
Grupo Bertrand, SA
Rua das Terras dos Vales, 4-A
Apartado 37
P-2700 Amadora Codex
Tel. (01) 49 59 050
Telex 15798 BERDIS
Fax 49 60 255

UNITED KINGDOM

HMSO Books (Agency section)
HMSO Publications Centre
51 Nine Elms Lane
London SW8 5DR
Tel. (071) 873 9090
Fax 873 8463
Telex 29 71 138

ÖSTERREICH

**Manz'sche Verlags-
und Universitätsbuchhandlung**
Kohlmarkt 16
A-1014 Wien
Tel. (0222) 531 61-0
Telex 112 500 BOX A
Fax (0222) 531 61-39

SUOMI

Akateeminen Kirjakauppa
Keskuskatu 1
PO Box 128
SF-00101 Helsinki
Tel. (0) 121 41
Fax (0) 121 44 41

NORGE

Narvesen information center
Bertrand Narvesens vei 2
PO Box 6125 Etterstad
N-0602 Oslo 6
Tel. (2) 57 33 00
Telex 79668 NIC N
Fax (2) 68 19 01

SVERIGE

BTJ
Tryck Traktorvägen 13
S-222 60 Lund
Tel. (046) 18 00 00
Fax (046) 18 01 25

SCHWEIZ / SUISSE / SVIZZERA

OSEC
Stampfenbachstraße 85
CH-8035 Zürich
Tel. (01) 365 54 49
Fax (01) 365 54 11

CESKOSLOVENSKO

NIS
Havelkova 22
13000 Praha 3
Tel. (02) 235 84 46
Fax 42-2-264775

MAGYARORSZÁG

Euro-Info-Service
Pf. 1271
H-1464 Budapest
Tel./Fax (1) 111 60 61/111 62 16

POLSKA

Business Foundation
ul. Krucza 38/42
00-512 Warszawa
Tel. (22) 21 99 93, 628-28-82
International Fax&Phone
(0-39) 12-00-77

ROUMANIE

Euromedia
65, Strada Dionisie Lupu
70184 Bucuresti
Tel./Fax 0 12 96 46

BULGARIE

D.J.B.
59, bd Vitocha
1000 Sofia
Tel./Fax 2 810158

RUSSIA

**CCEC (Centre for Cooperation with
the European Communities)**
9, Prospekt 60-let Oktyabria
117312 Moscow
Tel. 095 135 52 87
Fax 095 420 21 44

CYPRUS

**Cyprus Chamber of Commerce and
Industry**
Chamber Building
38 Grivas Dhigenis Ave
3 Deligiorgis Street
PO Box 1455
Nicosia
Tel. (2) 449500/462312
Fax (2) 458630

TÜRKIYE

**Pres Gazete Kitap Dergi
Pazarlama Dağıtım Ticaret ve sanayi
AŞ**
Narlibahçe Sokak N. 15
Istanbul-Cağaloğlu
Tel. (1) 520 92 96 - 528 55 66
Fax 520 64 57
Telex 23822 DSVO-TR

ISRAEL

ROY International
PO Box 13056
41 Mishmar Hayarden Street
Tel Aviv 61130
Tel. 3 496 108
Fax 3 544 60 39

CANADA

Renouf Publishing Co. Ltd
Mail orders — Head Office:
1294 Algoma Road
Ottawa, Ontario K1B 3W8
Tel. (613) 741 43 33
Fax (613) 741 54 39
Telex 0534783

Ottawa Store:
61 Sparks Street
Tel. (613) 238 89 85

Toronto Store:
211 Yonge Street
Tel. (416) 363 31 71

UNITED STATES OF AMERICA

UNIPUB
4611-F Assembly Drive
Lanham, MD 20706-4391
Tel. Toll Free (800) 274 4888
Fax (301) 459 0056

AUSTRALIA

Hunter Publications
58A Gipps Street
Collingwood
Victoria 3066
Tel. (3) 417 5361
Fax (3) 419 7154

JAPAN

Kinokuniya Company Ltd
17-7 Shinjuku 3-Chome
Shinjuku-ku
Tokyo 160-91
Tel. (03) 3439-0121

Journal Department
PO Box 55 Chitose
Tokyo 156
Tel. (03) 3439-0124

SINGAPORE

Legal Library Services Ltd
STK Agency
Robinson Road
PO Box 1817
Singapore 9036

AUTRES PAYS
OTHER COUNTRIES
ANDERE LÄNDER

**Office des publications officielles
des Communautés européennes**
2, rue Mercier
L-2985 Luxembourg
Tél. 499 28 1
Télex PUBOF LU 1324 b
Fax 48 85 73/48 68 17

10/92

Bibliothèques	Libraries
Université d'Ottawa	University of Ottawa
Echéance	Date Due

05 DEC. 1994

08 DEC. 1994

04 AVR. 1995

3 AVR. 1995

NOV 1 2 1995

NOV 05 1995

19 JUIN 1996

19 JUIN 1996

JUN 2 7 1996